conran DESIGN *guides*

TABLEWARE

JEREMY MYERSON
& SYLVIA KATZ

VNR VAN NOSTRAND REINHOLD
New York

Series Editor Joanna Bradshaw
Editor Mary Davies
Editorial Assistant Sally Poole
Design Paul Welti
Illustrator Cherrill Parris
Picture Research Nadine Bazar
Picture Research
Assistant Gareth Jones
Production Sonya Sibbons

Library of Congress Cataloging-in-
Publication Data

Conran, Terence.
    [Design Guides]
    The Terence Conran design
guides : tableware / edited by
Terence Conran
        p. cm.
    ISBN 0-442-30293-2
    1. Conran, Terence.   2.
Tableware–Themes, motives.   I.
Title.
    NK8725.5.C67A4   1990
    642'.7–dc20   89-70680
                                CIP

Published in the United States of
America by
Van Nostrand Reinhold
115 Fifth Avenue
New York, New York 10003

First published in Great Britain in
1990 by
Conran Octopus Limited

Typeset by
Servis Filmsetting Limited
Printed by Wing King Tong

ACKNOWLEDGMENTS

The publisher thanks the following
photographers and organizations
for their kind permission to
reproduce the photographs in this
book:
2–3 By Courtesy of the Board of
Trustees of the Victoria and Albert
Museum; 5 above centre Bodum; 5
below centre Kosta Boda; 5
bottom Katz Collection; 6 Ann
Ronan Picture Library; 9 Prudence
Cuming Associates, courtesy
Fischer Fine Arts Ltd, London; 10
right courtesy Design Museum; 10
left Josiah Wedgwood & Sons Ltd,
Barlaston, Stoke-on-Trent; 13 DP
Milan; 14 courtesy Design Museum;
16 Christie's Colour Library; 19
Kosta Boda; 24 left Conran Design
Group; 24 right courtesy Design
Museum; 25 Conran Design Group;
26 Oggetti, London; 26–27
Sotheby's London; 27 Arzberg
Porcelain Company; 28 By
Courtesy of the Board of Trustees
of the Victoria and Albert Museum;
28 below photograph by Courtesy
Frisse – courtesy Gallery
Association of New York State,
Inc; 29 above Sotheby's London;
30 Private Collection of Oy
Wartsila Ab Arabia, Finland; 31
Oggetti, London; 32 Royal
Copenhagen; 33 above Rosenthal;
33 below By Courtesy of the Board
of Trustees of the Victoria and
Albert Museum; 34 above
Collection du Musée des Arts
Decoratifs, Paris; 34 centre Studio
Sowden; 34 below Studio Sowden;
35 above Swid Powell; 35 below
David Mellor; 36–37 Bodum; 38
above Reunion des Musées
Nationaux; 38 below Angelo
Hornak; 39 Angelo Hornak; 40
above Royal Copenhagen; 40
below Germanisches
Nationalmuseum, Nurnberg; 41
Bauhaus-Archiv, Berlin; 42 above
Virginia Museum of Fine Arts, Gift
of Sydney and Frances Lewis; 42
below photograph by Courtney
Frisse – courtesy Gallery
Association of New York State,
Inc; 43 below Christie's Colour
Library; 44 Christie's Colour
Library; 44–45 Stelton of
Denmark; 45 left courtesy Design
Museum; 45 right DP Milan; 46
Bodum; 47 DP Milan; 47 below
Aldo Ballo; 48–49 Kosta Boda; 50
Department of History of Art/
University of Manchester (The
National Trust, Wightwick Manor);
50–51 Museum of Decorative Arts,
Prague (photograph by Gabriel
Urbanek); 51 Schott; 52 above
Artek Oy; 52 below Angelo
Hornak; 53 right Orrefors; 54
above Dartington Crystal; 54
below Glassexport; 55 Kosta Boda;
56 Sottsass Associati; 57 above
Matteo Thun; 57 below Carlo
Moretti; 59 above Cristalleries
Baccarat; 59 below Aldo Ballo; 60–
61 Katz Collection; 62 Virginia
Museum of Fine Arts, Gift of
Sydney and Frances Lewis; 63 left
Royal Copenhagen; 63 right
Hunterian Art Gallery/University
of Glasgow, Mackintosh Collection;
64 left Henning Christoph; 64
right Royal Copenhagen; 65 above
Katz Collection; 65 below David
Mellor; 66 above Rosenthal; 66
below Robert Welch; 67 above
Aldo Ballo; 67 below David
Harman Powell; 68 above Aldo
Ballo; 68 below Clive Corless/
Conran Octopus; 69 above Forma
House Ltd; 69 below Kartell; 70
RFSU Rehab; 70–71 David
Mercatali; 72 Artek Oy; 73 DP
Milan; 74 Topham Picture Library;
75 Stelton of Denmark; 77
Archivio Gio Ponti (photograph by
Gasparini); 79 Sottsass Associati
(photograph by Gitty Darugar); 78
Bodum.

The following photographs were taken specially for Conran Octopus by Simon Lee:

5 top, 22–23, 29 below, 30–31, 43 above, 53 left, 58, 71.

We would like to thank the following for their cooperation:
The Conran Shop
Design Museum
Rosenthal
Sasaki

Every effort has been made to trace the copyright holders and we apologize in advance for any unintentional omission and would be pleased to insert the appropriate acknowledgment in any subsequent edition of this publication.

**AUTHORS'
ACKNOWLEDGMENTS**
The authors wish to thank all those manufacturers and designers who answered queries and searched through their archives, the supportive and professional staff at Conran Octopus and Sir Terence Conran for his personal interest and guidance.

NOTE TO READER
Names of objects and designers printed in roman or **bold** type denote that a photograph of the object or a biography of the designer can be found elsewhere in the book.

THE SIGNIFICANCE OF TABLEWARE DESIGN

The way we eat and entertain, and the objects we put on our tables relate to the most fundamental cultural rituals of our society. Over centuries the rite of dining has evolved into a complex social activity in which rules are obeyed and customs observed. Table manners and table objects say a lot about us – the cultures to which we belong, the economic systems to which we adhere and our sense of our own style and status.

Victorian silver-plate *epergne*: cultural values of age reflected in the ritual of dining.

Tableware design also reveals the technology of an age, the industrial processes which have been developed and the materials which are in vogue or, simply, available and affordable. Above all, in the twentieth century tableware presents the designer with a unique challenge: given the industrialization and relative democratization of the market-place, it is in many ways the ultimate mass-market commodity.

Every home in the modern world needs such basic items as crockery, cutlery and glassware, and for that reason, the challenge of designing tableware has attracted some of the greatest architects and designers, artists and sculptors of the twentieth century.

It is true that the medium offers great opportunities to explore creative expression, materials technology and cultural change. But the overriding impulse has been the simple desire to influence popular taste – an irresistible opportunity, especially for those designers driven by social and moral imperatives, to make a significant contribution to the improvement of material life.

But it is not just great designers and architects who have been associated with tableware. Some of the most enlightened and progressive industrial companies have prospered in a field which requires continual refinement of production techniques to complement the delicate balance of function and aesthetics in the area of design.

A number of manufacturers have at different times dominated the story of twentieth-century tableware due to their technical versatility and sympathetic patronage of leading designers. Their names will be familiar to anyone interested in good design for the home: Alessi of Italy, Arabia of Finland, Orrefors of Sweden, Rosenthal of Germany, Wedgwood of Britain, Georg Jensen of Denmark, and there are many others.

Their products are characterized by a sense of national identity, coupled with some unique quality which differentiates them from their competitors. They have contributed not only to the evolution of design but to the progress of modern manufacturing.

This book celebrates the output of these companies and the work of the designers they have commissioned. But it is more than just a guide to the outstanding and influential tableware designs of the past hundred years. By highlighting the conditions in which they were created, it comments on the dramatic cultural, technological and aesthetic changes which have taken place in the home during this century.

THE RELATIONSHIP OF ART, CRAFT AND INDUSTRY

Twentieth-century tableware design has combined the provision of ranges for the mass market with the creation of more expensive, élitist items. That distinction is not new. Before the Industrial Revolution introduced new possibilities for mechanization, the agrarian poor depended on rudimentary utensils and equipment made by local craftsmen for use at table. But for the rich and titled and the expanding bourgeoisie it was an entirely different story. The complicated ritual of reflecting status through dining was in play and much notable work emerged in the process.

The Industrial Revolution in the early nineteenth century had enabled mass-consumption tableware to be machine produced, but the first machine-made products continued to imitate handmade ones, except that they lacked the human qualities of genuine craft work.

It was the Arts and Crafts movement founded in Britain by William Morris (1834–96) which voiced the most violent opposition to factory production, regarding it as degrading to the human spirit, and set about gaining support for the concept of wholesome craftsmanship. The movement, which featured such prominent names as the design theorists C R Ashbee and C F A Voysey, revived the medieval craft guilds.

However, this defiant stand against the march of progress left a confused design legacy. The Arts and Crafts movement profoundly influenced the Art Nouveau designers and Vienna Secessionists who in the early years of the twentieth century were laying the groundwork for modern design with its glorification of the machine and urban life.

In particular, the Wiener Werkstätte founded by Secessionists Josef Hoffmann and Kolo Moser was based on C R Ashbee's Guild of Handicrafts. Yet it was at the Wiener Werkstätte that Hoffmann created important metalware

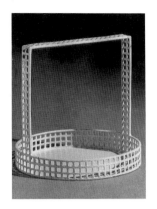

Painted metal tray by Josef
Hoffmann, 1905: practical
geometry prefigures
Modernism.

French Apilco cup and saucer:
timeless classic endures
through an age of industrial
advance.

designs for cutlery and trays which prefigured in their simple
beauty and repeating geometry the modern era of practical,
stylish, utilitarian tableware lines.

The Arts and Crafts movement also influenced the founding
in 1907 of the Deutscher Werkbund, an association aimed at
uniting art and industry in a focus on the function of objects. A
prime mover in the Deutscher Werkbund was Peter Behrens,
artistic adviser to the German industrial giant AEG
(Allgemeine Elektrizitäts-Gesellschaft), who was committed to
developing a new aesthetic and improved quality for industrial
production.

Groups such as the Wiener Werkstätte and the Deutscher
Werkbund influenced the famous Bauhaus design school
founded in Germany in 1919, and effectively set a course for
mass-commodity utilitarian tableware production in twentieth-
century Europe. Meanwhile in the USA industry had
mechanized rapidly in a straightforward, pragmatic way,
uninhibited by the moral ambiguities raised by Morris and
those he influenced.

Throughout this century then, craft and factory skills have
coexisted, interchanging ideas and approaches. At different
times hand techniques have become possible by machine, and
new materials have emerged to stimulate radical change. The
process continues today, and an intricate interweaving of art,
craft and industry has emerged in tableware production.

FOUR GIANTS OF TWENTIETH-CENTURY TABLEWARE

*The origins of four European companies in particular –
Wedgwood, Rosenthal, Georg Jensen and Alessi – point to the
evolution and nature of twentieth-century tableware design.
Each has brought a special dimension to the market-place;
each represents a particular transition in design; and,
significantly, each has enjoyed success in the USA.*

WEDGWOOD

*Wedgwood, one of the world's leading names in ceramics, was
founded in 1759 in Staffordshire, England, by Josiah
Wedgwood, a master potter. It was the first company to mass-
produce goods at the start of the Industrial Revolution,
pioneering new production techniques and the division of
labour; and thereby anticipating Henry Ford's model of
manufacture by 150 years. It even developed a design*

The legacy of Wedgwood:
beauty and practicality in
design, epitomized by the
Queensware line, and
pioneering moves towards
modular, standardized ranges
for production.

*vocabulary of basic geometric shapes, thus prefiguring
Modernist design principles.*

*Josiah Wedgwood liked simple, practical design – 'useful
ware' – and created the core ranges himself, including the
Queensware line, so called after he became Potter to Queen
Charlotte in 1765. Wedgwood clearly directed his products*

towards a new consumer group, the expanding middle classes, using intelligent marketing strategies to exploit late eighteenth-century fashion. For example, his famous Black basalt teapot was produced to show off bleached white hands, in vogue among women at the time.

But Josiah Wedgwood was not interested in the central tenet of mass production: to sell as many products as cheaply as possible. He remained committed to the idea of high-quality objects for taste-conscious customers, commissioning leading fine artists of the day on certain ranges. The most important of these was John Flaxman, who was the first true artist-designer.

It is with a tradition of class and quality that Wedgwood has traded in the twentieth century. The Wedgwood name has not only endured – although the company is now owned by the Irish crystal-maker Waterford – but the founder's legacy of standardized design for production is still with us.

ROSENTHAL

The German manufacturer Rosenthal is not as long established as Wedgwood but it has produced arguably the most significant ceramic tableware of the twentieth century. The company was founded by Philipp Rosenthal in the dungeons of Erkersreuth Castle in 1879.

Philipp Rosenthal had already visited the USA and noted the ready-market there for simple, well-defined designs with careful decoration. From the start, exports were a feature of Rosenthal's sales strategy but the real secret of the company's success was its ability to absorb new ideas from a vortex of radical change in art and design at the turn of the century.

While Philipp Rosenthal developed a strong industrial base, building a new factory in 1897 and then taking over several companies including the Thomas china company between 1908 and 1928, he remained aware of the formal revolt by artists and architects against nineteenth-century design tradition.

Kurve cutlery by Tapio Wirkkala, late 1960s: Rosenthal's encouragement for sculptural expression.

Rosenthal decoration for early tableware: Darmstadt set, 1904, gives artistry to simple shape.

The formal quality and line of Rosenthal products and their respect for materials appeared to be in the vanguard of the Modern movement. But Philipp Rosenthal had an uneasy relationship with Bauhaus founder Walter Gropius. In particular they clashed over whether table articles needed decoration – Rosenthal considered they did. Later, after 1945, the company and Gropius were reconciled. Gropius became a trusted adviser and important designer for Rosenthal. His TAC I tea set of 1968 is one of the great tableware classics of the twentieth century.

But Gropius was by no means the only leading designer associated with Rosenthal. In the 1950s Philipp Rosenthal Jr initiated a radical change of style combining functional form with aesthetically progressive design. As a result, in 1954, the company launched the Rosenthal Line, later known as the Rosenthal Studio-Line, with the Form 2000 service designed by Raymond Loewy and Richard Latham. Finnish Modern master Tapio Wirkkala created the Century Service, which commemorated the company's centenary in 1979. And many other designers and artists, including Timo Sarpaneva, Nick Roericht, Queensberry Hunt, Eduardo Paolozzi and Salvador Dali, have contributed to Rosenthal's success.

GEORG JENSEN SILVERSMITHY

Georg Jensen Silversmithy was founded in Copenhagen by a Danish sculptor-turned-silversmith in 1904. From humble beginnings in a tiny workshop Georg Jensen became one of the most famous silverware names in the world. He created a style which was accessible and inexpensive. His jewellery appealed to the middle classes unable to afford opulent stones, and his tableware designs – cutlery and teapots – enhanced his reputation as a sensitive creator of form. He never forgot his ambitions as a sculptor and made an important contribution to the natural elegance of the Scandinavian Modern movement.

Sterling silver jug by Sigvard Bernadotte, 1938: refined elegance for Georg Jensen.

Many leading artists, sculptors and designers have been associated with the Georg Jensen company over the years, among them the noted silver designer Johan Rohde with whom Jensen collaborated throughout his career, Henning Koppel, Kay Bojesen and Sigvard Bernadotte. Two of Jensen's sons, Jørgen and Søren Georg Jensen, joined the smithy as designers. Søren Georg, a sculptor by training, took up the artistic reins of the company after his father's death, and has maintained Georg Jensen's creative reputation by devoting much of his attention to exploring the cylindrical form of silver holloware.

A L E S S I

The Italian firm of Alessi is another distinguished European family-owned maker of metal products. It was founded by Giovanni Alessi in 1921 to make metal parts for other producers. But in 1924 Alessi began producing its own trays and coffeepots, and within three years had dedicated itself entirely to the manufacture of household items.

The materials used were mainly nickel-plated or silver-plated brass, and Alessi specialized in unusual items for the table, such as pepper mills, flask stands and cheese dishes.

From Alessi's Tea & Coffee Piazza Series: silver service by Modernist American architect Richard Meier.

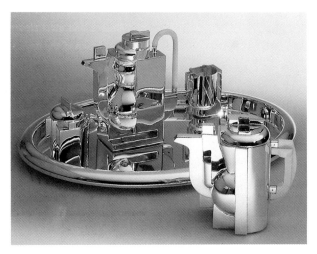

Tableware as mini-
architecture: Aldo Rossi's
design in Officina Alessi's 1983
collection.

After Giovanni's eldest son Carlo took over the firm in the early 1930s a distinctive Alessi style full of personality and strong images began to emerge, epitomized by his famous Bombé *service of 1945.*

When stainless steel supplanted silver-plated brass, collaborations with outside designers led to further commercial success in the 1950s. More recently the company has established a new trademark, Officina Alessi, *to market more experimental design projects free from the limits imposed by industrial mass-production. Under this name, 11 leading international architects produced in the early 1980s some of the most remarkable tea and coffee services seen this century – with shapes recalling the work of such modern design pioneers as Hoffmann and Puiforcat.*

DEVELOPMENTS IN MATERIALS AND TECHNOLOGY

Developments in design have consistently been linked with developments in materials. The introduction of bone china, porcelain, earthenware and stoneware have each, in turn, produced new shapes and forms.

What the progress of Wedgwood, Rosenthal, Georg Jensen and Alessi shows most clearly is that enlightened investment in the best design of the day creates markets rather than simply satisfying them. It also demonstrates that the consistent use of materials, whether porcelain in Josiah Wedgwood's time or silver in the case of Georg Jensen, is important for success.

Indeed developments in twentieth-century tableware are most clearly appreciated by considering the media in which designers have worked. For this reason the objects chosen for inclusion in this book are divided into four categories: china, metal, glass and plastics. Only cutlery, which has regularly mixed materials as the century has advanced, has a separate developmental path in the book (sharing the plastics section).

CHINA

Of all the materials used to make tableware, china is the oldest and most traditional. It has also had the most stable material technology in modern times with gradual and continual refinements of production technique throughout the twentieth century, but no important breakthroughs to which designers could respond.

Where there have been major changes is in firing and drying technology, enabling faster production and improved finishes, and in methods of applying decoration to ceramic bodies. Silkscreen printing was developed in the 1950s, as china tableware adopted the decorative technologies of other industries. In the late 1980s dust pressing, a technique formerly only used to make tiles, has developed enough sophistication to produce plates. This offers much potential for the future. But for much of this century, the traditional constraints on ceramics have remained, and the design classics which have emerged have been in response to what is essentially a timeless challenge.

Perhaps that is why so much of the best of twentieth-century china tableware has timeless appeal: Hermann Gretsch's *notable* 1382 *range for* Arzberg *of Germany:* Kaj Franck's Kilta *service of* 1952 *for Arabia of Finland; or* Russel Wright's *American Modern dinnerware produced by the Steubenville Pottery from* 1939 *to* 1959. *And perhaps that is also why early ranges by Wedgwood, Royal Copenhagen or Royal Doulton do not look out of place today.*

Of course china tableware has reflected the stylistic changes of the century, whether the vibrant Art Deco of Clarice Cliff *in the* 1930s *or the colourful Post-Modernism of Memphis or Swid Powell ceramics in the* 1980s. *But some pieces have pleasing forms of such simplicity and beauty, like the traditional* French Apilco *cup and saucer, that they are impossible to date. They will be as relevant in the next century as in this.*

METAL

Unlike the calm transitions in ceramic tableware, production of metalware and cutlery was revolutionized by the introduction of stainless steel. This was first developed in 1913 but it was not widely used until after World War Two. Domestic stainless steel, which has 18 per cent chromium and eight per cent nickel, not only encouraged a new 'lightweight' aesthetic of sleek lines, it abolished the need to electro-plate metals (coating them with silver by the process of electrolysis) in order to avoid harmful contact with food.

Much of the pioneering work in modern metalware had been electro-plated. British botanist Christopher Dresser, *widely recognized as the first industrial designer of the modern age, produced electro-plated teapots and soup tureens in the 1880s with a utilitarian styling which anticipated the Bauhaus movement by 40 years.*

The Bauhaus metalwork shop run by Laszlo Moholy-Nagy from 1923 to 1928 continued Dresser's experiments with functional tableware. One of its most famous designers, Marianne Brandt, produced an elegantly modern brass teapot in 1924 which was silver-plated inside to make it safe to use.

Two-handled bowl by Christopher Dresser for A Hukin and Heath, 1879: pioneering metalware needed to be electroplated.

Silver tea and coffee set by Jean Puiforcat for Elkington & Co, 1936–37: fine Art Deco.

Of course other tableware designers, such as the French master of Art Deco, Jean Puiforcat, Britain's Harold Stabler and Danish designers, Georg Jensen and Johan Rohde, worked directly in silver. But stainless steel changed the rules. Initially it was a shortage of silver during World War Two which forced manufacturers such as Jensen to use stainless steel, but very quickly the material became associated with the best of Scandinavian Modern design, and in the USA stainless-steel serving vessels developed a lightweight, sleek aesthetic which became very popular in the 1950s and 1960s.

In the 1970s, stainless steel suffered a loss of status as it became associated with cheap cafeterias. But in the 1980s, new ranges such as the Yamazaki Serving Collection of 1982 by

Coffeepot from the 1982 Yamazaki serving collection designed by Robert Welch: new direction for stainless-steel design.

British designer Robert Welch once more upgraded the image of the material as not only functional but elegant and pleasing to the eye.

GLASS

A series of nineteenth-century key advances has informed the direction of twentieth-century glass tableware. In 1827 mould pressing started in the USA with the use of machines, making cheap glass available on the mass market. The earliest known patent for pressed glass was for furniture knobs, but in the 1830s and 1840s many new designs emerged with the introduction of better moulds and mechanical presses. Then, in the 1860s, steam-powered presses were introduced, speeding up the production process.

The mid-nineteenth century also saw important developments in decorative glass: colour played a more significant part; acid etching was introduced using wax and hydrofluoric acid; and there was also a popular revival of cameo glass, a technique first discovered by the ancient Romans. Cameo glass was taken up in a spectacular way at the turn of the century, by Art Nouveau glass-makers, such as Emile Gallé and Louis Comfort Tiffany.

In the 1860s, pâte-de-verre (a technique, known in ancient Egypt, in which glass dough or paste is pressed into a mould to produce wonderful translucent effects) was rediscovered. It was first adopted commercially in 1906 and has been used to great effect by a number of leading artists during this century, most recently by British designer Diana Hobson.

In the 1880s, American glass-makers introduced metal moulds into which glass was blown by compressed air, allowing more control over the finished result. Famous uses include the Coca-Cola bottle and electric light bulbs. Then in 1915 came a further quantum leap: heat-resistant glass (known as Pyrex) was developed at the Corning Glassworks in the USA. Made from borosilicate glass, its slow expansion rate and good chemical stability enabled, for example, the production of glass tea services.

The significance of this advance can be seen most clearly in the fine dimensions (resembling laboratory test tubes) of heat-resistant glass tea services designed by Wilhelm Wagenfeld and Ladislav Sutnar in the early 1930s.

Glass tumblers by the legendary Finnish architect Alvar Aalto for the Iittala Glassworks, 1933.

Octave series by Bertil Vallien, 1977: cool simplicity from Kosta Boda of Sweden.

The technological parameters for twentieth-century table glass had been set. However, some designers, such as Italy's Paolo Venini, head of the Venini glassworks and creator of the famous Fazzoletti *vase in 1954, revived techniques used in Venetian glass centuries before.*

In 1962 there was a further dramatic change. Harvey Littleton, a chemist, ceramicist and teacher, combined with other glass technicians to form a group sponsored by the Toledo Museum of Art. Together they worked on a molten glass that would melt at a low temperature and devised a small furnace that could be operated by bottled propane gas. These developments enabled designers to create their own glass working solo, an event widely hailed as the most signficant step since the introduction of pressed-glass manufacture in the early nineteenth century.

It led to the New Wave Studio Glass Movement as designers and artists exploited their freedom of manufacture. An extraordinary outpouring of exciting and imaginative new ideas in table glass occurred in the 1970s and 1980s and the pace shows no signs of slackening.

PLASTICS
Of all the materials used to make modern tableware, only plastics can claim to be an entirely twentieth-century

phenomenon. The earliest plastics in use – Celluloid and Bakelite – were not considered suitable for products in which moisture, taste and smell were considerations. It was only when Beetle moulding powders were introduced for light-coloured resins in the mid-1920s that designers at last had a new medium in which to produce tableware. The results included the famous Bandalasta picnic sets and vacuum flasks of the late 1920s and early 1930s.

Then, in 1932, an improved material called urea formaldehyde arrived, destined to supplant the mottled colours of Bandalasta in the homes of the 1930s. But the big breakthrough came in 1939, when the American Cyanamid Company made melamine commercially available for the first time. After World War Two melamine was widely used by tableware manufacturers, its colour, lightness, cheapness and durability making it highly suitable for domestic use.

The first melamine tableware was designed by the American Russel Wright for Cyanamid in 1949 under the name Meladur. But although this was used by a New York restaurant chain it never found favour in the home. Then, in 1953 Wright created the Residential dinnerware range in melamine for the Northern Industrial Chemical Moulding Company of Boston: this achieved a breakthrough in domestic acceptance and was widely imitated.

In Britain the Melaware range of 1959 designed by A H Woodfull and the Fiesta range of 1961 designed by Ronald Brookes demonstrated the versatility of melamine. But it was the Italians who seized on the new possibilities in plastics tableware. Plastics enabled a more freeform use of shapes and encouraged stackability and inter-lockability. A new design vocabulary emerged, typified by Enzo Mari's Pago-Pago vase for Danese in 1969, which could be used either way up, and by Massimo Vignelli's Max I melamine tableware for an American company, Heller, in 1972.

Teacup from Russel Wright's Residential dinnerware, 1953: a breakthrough for melamine tableware in the USA.

Picboll by Guzzini, 1978:
Italians explore the creative
and technical versatility of
modern plastics.

Plastic has proved to be the most inventive and fastest-changing tableware material this century. It has influenced the development of cutlery, replacing more traditional materials on the handles of ranges by David Mellor of Britain and Andrée Putman of France. It has also been in the forefront of developments in dishwasher-proof and microwave-proof ware in the 1980s.

THE FUTURE

Throughout this century designers have explored their own powers of visualization and imagination in the context of technical innovation and cultural change. This process continues today as designers face up to new challenges in tableware design. Flexible manufacturing systems now offer high-tech ways to modify factory production runs to enable subtle design variations. More travel and increasingly cosmopolitan lifestyles mix different cultures from all over the world. Already East and West are exchanging ideas on cuisine and social codes. Such trends will be accentuated in the next few years and exciting new ideas in tableware are bound to emerge as a result.

Lacquered soya plate and
chopsticks for Kushoda of
Japan by Toshiyuki Kita:
Eastern cuisine an influence
for change.

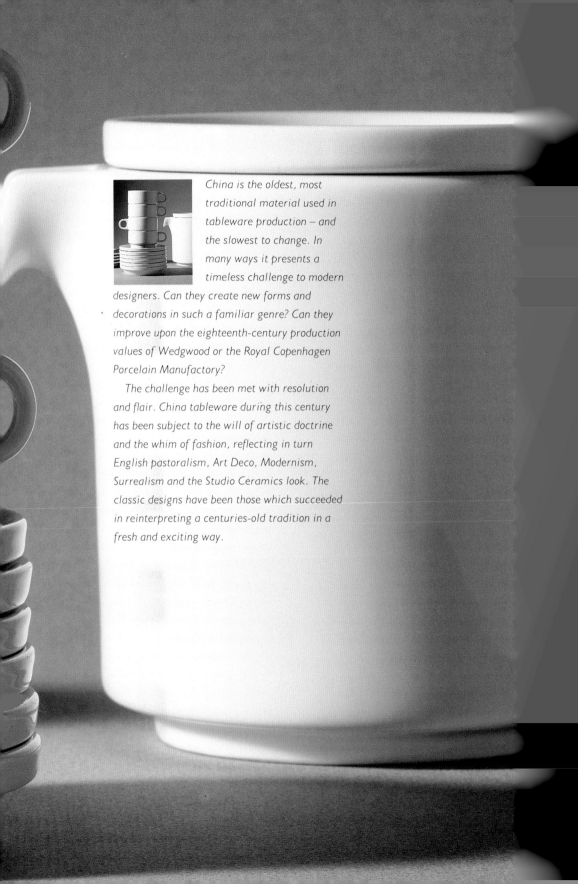

China is the oldest, most traditional material used in tableware production – and the slowest to change. In many ways it presents a timeless challenge to modern designers. Can they create new forms and decorations in such a familiar genre? Can they improve upon the eighteenth-century production values of Wedgwood or the Royal Copenhagen Porcelain Manufactory?

The challenge has been met with resolution and flair. China tableware during this century has been subject to the will of artistic doctrine and the whim of fashion, reflecting in turn English pastoralism, Art Deco, Modernism, Surrealism and the Studio Ceramics look. The classic designs have been those which succeeded in reinterpreting a centuries-old tradition in a fresh and exciting way.

BROWN BETTY TEAPOT
Price & Kensington Potteries
c. 1890
Made in Stoke-on-Trent, the capital of the English pottery towns, this popular classic has not changed since it was first designed. The Brown Betty is a masterly piece of functional design which comes in a variety of sizes. Its lid, which has a hole to allow steam to escape, remains secure when even the last drop of tea is being poured; the large, comfortable handle keeps the hand away from the heat of the pot. The arrangement of its parts enables the maximum amount of tea to be easily lifted and the functional parameters of the design have been combined to create a pleasing shape. When it was first launched, the Brown Betty was such a success that Price & Kensington turned its entire factory, which till then had produced a range of pottery products, to teapot manufacture.

BLACK BASALT TEAPOT
Josiah Wedgwood for Josiah Wedgwood & Sons 1768
A design which has been in production for more than 200 years, it does not look dated. Made in an unglazed black stoneware which is rich, tough and matt, this stable and stylish teapot has a timeless beauty. Wedgwood had a natural instinct for interpreting the fashions and whims of his era.

ENGLISH SCENIC TABLEWARE
William Adams 1890

This tableware range developed by a former Wedgwood apprentice is very English and very traditional. It features pastoral scenes of country life in hand-engraved prints, which have always been an English speciality. The decoration is printed under the glaze to create a durable finish. Adams's English Scenic enjoyed a revival with the country-house set in the early 1970s as rustic lifestyles took hold – at least over the weekend. Today the range still looks best adorning an old pine country dresser.

CORNISHWARE
T G Green & Co 1920s

The distinctive range of blue and white striped tableware began with storage jars and jugs – basic utilitarian objects. But over the years its bold, simple character has attained the status of a high-style classic sold in the top designer stores. The range has also been extended to include the full collection of tableware products. Today it is freezer-, microwave- and dishwasher-proof – a sophisticated departure from its basic origins.

BIZARRE POTTERY
**Clarice Cliff
for A J Wilkinson** 1929

This is the tableware range which enjoyed such instant acclaim that Clarice Cliff became Britain's most popular Art Deco designer of the 1930s. So great was the demand for the hand-painted Bizarre collection that Cliff was forced to hire young painters to follow her style. Its brilliant colours and geometric shapes show a Cubist influence.

MODEL 1382
Hermann Gretsch
for Arzberg 1931

Widely regarded as one of the greatest achievements in twentieth-century tableware, this white porcelain design is the first truly Modern range. German designer Hermann Gretsch was not a Bauhaus member but the classic simplicity of his work for Arzberg embodied the virtues of the Modern movement in Germany. Its sensible, easy-to-grip handles and elegant, practical shape with pleasing curves have been much emulated but rarely bettered. Model 1382 won design awards in Milan in 1936 and also in Paris in 1937.

KESTREL COFFEE SET
Susie Cooper
for Susie Cooper Pottery
1932–5
Like Clarice Cliff, the work of Susie Cooper is rooted in the artistic traditions of the Potteries in England. Her Kestrel shape was made by Wood & Sons and decorated in her Burslem studio. It furthers Cliff's geometric style, but Cooper was also a great individualist: Kestrel has a complex surface pattern with *sgraffito* banding.

AMERICAN MODERN
Russel Wright
for Steubenville Pottery
1939
The greatest artistic and commercial success of the USA's best-known tableware designer of the 1940s and 1950s, American Modern remained in production for 20 years. The natural grace and purity of the design revealed Russel Wright's Midwestern Quaker roots and training as a sculptor. The pieces, which are now collector's items, have a flowing quality which escapes stylistic labels. Organic in form, they were first produced in white but later glazed with colours.

EARTHENWARE MUGS AND PITCHER
Eric Ravilious
for Josiah Wedgwood & Sons

The British artist and designer Eric Ravilious (1903–52) worked extensively for Wedgwood from 1935 and the company continued to release his designs after his death. The pitcher (*centre*) and Alphabet mug (*right*) belong to the late 1930s whereas the Coronation mug was launched in 1953. Ravilious was a wood engraver and watercolourist, which explains his delightfully detailed light touch and subtle use of colour.

STACKING SERVICE
Nick Roericht
for Rosenthal 1959

Each unit in this space-saving German porcelain collection has been designed to fit perfectly with its fellows, anticipating an age of living in which storage space would be at a premium. Roericht's pursuit of stackability created dual-function vessels (his soup bowls have spouts for serving and pouring), but also a clarity and beauty in appearance.

FORM 2000
Raymond Loewy and Richard Latham
for Rosenthal 1954

This service launched German manufacturer Rosenthal's Studio-Line, which was dedicated to creating original works with a high artistic content. Its self-conscious, highly stylized look – each piece has a waist – is typical of the showman's flair evident in American design pioneer Raymond Loewy's best work. Its angularity and elongated form are also typical of work produced in the 1950s.

KILTA SERVICE
Kaj Franck
for Arabia 1952

Franck's brightly-coloured utility Kilta earthenware service is one of the classic objects of Finnish design: inexpensive, accessible, enduring and popular. The name Kilta means 'guild', a reference to the associations of craftsmen formed in medieval times, and Franck sought to convey a handmade, studio feel in a factory-manufactured product. The multi-purpose Kilta enabled customers to build up a collection piece by piece. It sold in great numbers in Finland and has been reissued.

DECORATIVE PLATES
Piero Fornasetti

c. 1950

Fornasetti's background in the theatre and in Surrealist painting is evident in these transfer-printed ceramics. One of the most unusual and idiosyncratic talents on the Italian post-war design scene, Piero Fornasetti (1913–88) applied highly personal graphic designs to dinner plates and furniture. These pieces are intriguing: they appear to be fragments suggesting a larger picture left to the viewer's imagination. Fornasetti's work enjoyed a revival of interest in the 1980s as the idea of surface ornament in design was reassessed.

TAC I TEA SET
Walter Gropius
for Rosenthal 1962
One of the most celebrated editions of Rosenthal's Studio-Line and one of the most famous china shapes of the century, this service by the founder of the Bauhaus is an essay in sculpture which combines beauty and utility. It is a lasting demonstration of Gropius's radical instinct in design: all aspects of the function have been examined anew. The form and positioning of handle and spout enable safer pouring while giving the service its special graphic personality. It was Gropius's triumphant assertion that Modernism had not run its course by the 1960s.

BLUE LINE
Grethe Meyer
for Royal Copenhagen Porcelain Manufactory 1964–5
Industry and art have been combined effectively at Denmark's most prestigious manufacturing company, the Royal Copenhagen, ever since it was established in 1775. But Meyer's Blue Line earthenware dinner set can be seen as representing a significant turning point for the company: its modernity, simplicity and stackability brought a rich tradition of tableware manufacture up to date, and the reasonable price made it more accessible.

CONCEPT
Martin Hunt
for Hornsea Pottery 1977

Tableware manufacture grew in sophistication as Martin Hunt of the Queensberry Hunt design partnership collaborated with Colin Rawson of Hornsea Pottery to create an imaginative English range which utilized fully vitrified clay. This is a very tough and non-porous material. Concept's decoration is innovatively integrated into the form via sculpted contours rather than applied as a surface pattern.

CENTURY SERVICE
**Tapio Wirkkala
for Rosenthal** 1979

Specially designed by one of the greatest names of the Scandinavian Modern movement to celebrate Rosenthal's centenary, this service combines artistic originality and technical innovation. Wirkkala, artistic consultant to Rosenthal's Studio-Line, evoked the shell of a sea urchin in his natural, expressive design. Rosenthal's production techniques ensured a delicate texture which enhanced the service's rounded quality.

DECORATED CERAMICS
George Sowden and Nathalie du Pasquier 1986–7

British designer George Sowden and his French partner Nathalie du Pasquier, two of the leading figures in the Milan-based Memphis movement, consistently and exuberantly explored the potential of surface pattern during the 1980s. These painted ceramic one-offs show their penchant for drawing inspiration from a wide range of historical and stylistic sources.

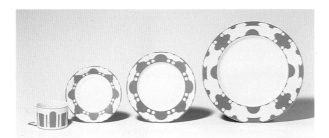

GLAZED STONEWARE POTTERY
**Janice Tchalenko
for Dartington Pottery**
1986
The British Studio Pottery look of the 1980s is epitomized by the bold colour and domestic shapes of Tchalenko's work. This collection, which features 19 objects in a range of four patterns, has a handmade feel even though it is mass moulded by Dartington. It is a world away from the technical artistry of Loewy or Gropius for Rosenthal yet has its own charm and integrity.

VERONA
Stanley Tigerman and Margaret McCurry for Swid Powell 1987
New York tableware maker Swid Powell invited leading American architects to decorate the company's wares for the Cityline series with some success. Stanley Tigerman's Verona collection employs the theme of an aerial view of an ancient walled city. The use of designer names to decorate china marks its transition from utilitarian object to high-style possession.

The most influential development in modern metal tableware has been the widespread introduction of stainless steel after 1945. It obviated the need to electroplate metals in contact with food and created a new aesthetic associated with the finest examples of fluent Scandinavian Modernism.

But a century of distinctive metal tableware production was effectively summed up in the early 1980s when Alessi commissioned the Piazza programme, a series of outstanding tea and coffee services from 11 of the world's leading contemporary architects.

SOUP TUREEN
Christopher Dresser
1880

Dresser's design demonstrates an interest in utilitarian styling which prefigured the concerns of the Bauhaus by 40 years. The soup tureen stands on three legs and has ivory handles and knob. It is electroplated to enable safe contact between metal and food, a technique later superseded by the invention of stainless steel. Dresser practised as an industrial designer a full half-century before the modern industrial-design profession was established.

TUDRIC TEA-SET
Archibald Knox
for Liberty & Co. c. 1903

Knox was Liberty's most prolific silver designer and this popular service illustrates the influence of Celtic art on his work. It was so successful in silver that it was reissued in pewter to reach a wider market. The relief patterning, enamel insets and wickerwork handles evoke the Art Nouveau style.

TEA SERVICE

Josef Hoffman c. 1909

This classic design in brass with wooden handles is inscribed with the linked WW emblem of the Wiener Werkstätte. Josef Hoffman was a founder in 1903 of these self-styled Vienna Workshops, which were dedicated to exploring new principles in modern design. Six years earlier he had joined a radical group of artists, architects and designers to form the Vienna Secession, a counterblast to traditional Austrian design. Hoffman's work has thus had an important bearing on the development of Modernism in Europe. His simple, geometric interpretation of Art Nouveau, as seen in this tea service's angular character and regular planes, has been widely regarded as laying the foundations of Art Deco. This piece was designed at the same time as Hoffman was working on his architectural masterwork, the magnificent Palais Stoclet in Brussels.

JUG
Johan Rohde
for Georg Jensen 1920

The sculptural and organic tradition in Danish modern design is epitomized by the silver craftsman Johan Rohde, who was chief designer at the famous Georg Jensen Silversmithy in Copenhagen. This silver jug has a flowing grace and beauty which suggests why Jensen's industrial output had such a profound influence on Scandinavian Modernism.

COPPER TEAPOT
Marianne Brandt
for the Bauhaus 1924

Brandt was one of the most famous designers to emerge from the Bauhaus. This smaller copper teapot was developed in the metalwork shop run by Laszlo Moholy-Nagy. It is silver-plated inside with a silver tea-infuser and an ebony handle. Its structural composition has been very influential even though the piece is diminutive: it stands only 8 cm (3 in) high.

TEA MACHINE
**Wolfgang Tumpel
for the Bauhaus** 1927

Tumpel's electroplated tea kettle, known as the *Teemaschine*, is a fascinating demonstration of Bauhaus principles. Made by hand in the metalwork shop after the Bauhaus had moved from Weimar to Dessau, it reflects the form-follows-function teaching of Walter Gropius and his colleagues. Its aesthetic derives entirely from the manufacturing considerations of producing an electric plug-in machine for making tea. The tea-infuser is held in a small cylinder, unusually off-centre, which is attached to the main body of the product and becomes a functional feature in its own right. In being honest to its materials and industrial origins, Tumpel's design has an appearance which achieves the same degree of presence and personality as more consciously styled and decorated tableware objects of the 1920s.

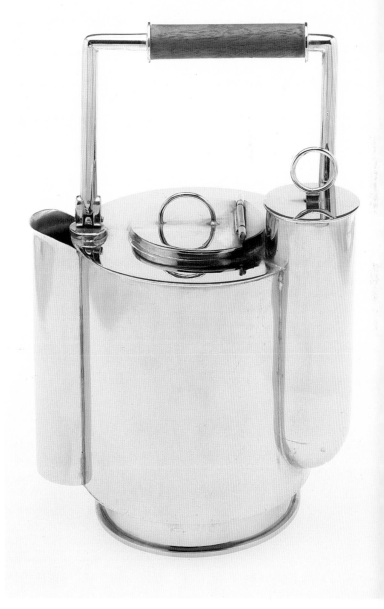

TEA AND COFFEE SERVICE
Jean Puiforcat 1937

This silver service with rosewood handles is typical of the emphatically graphic style of French sculptor-silversmith Jean Puiforcat, whose Art Deco designs were highly popular in England and France during the 1920s and 1930s. Puiforcat's approach was based on the harmony and geometry of the Ancient Greeks, yet the smooth, elitist quality of his work also prefigures Alessi's architect-designed tea and coffee services of the 1980s.

SPUN-ALUMINIUM TEA SERVICE
Russel Wright early 1930s

An early range by the USA's most influential mid-century tableware designer. The stylish practicality of the spun-aluminium of this design brought aluminium out of the kitchen and encouraged the idea of more informal entertaining.

BOMBE LINE
Carlo Alessi
for Alessi 1945
Bombé means bloated and
exaggerated. This milestone
tea and coffee service in
stainless steel exuberantly
sets aside classical harmony in
favour of design personality
and vigour. Bombé changed
the entire nature of tableware
and came to symbolize
Alessi's strong and successful
emphasis on sophisticated
design imagery after 1945.

FISH DISH
Henning Koppel
for Georg Jensen 1954
Henning Koppel trained as a
sculptor as well as a designer
in Copenhagen and Paris, and
his fluid organic style is most
closely associated with the
finest work produced by the
Georg Jensen Silversmithy
after 1945. This silver fish dish
is typical of Koppel's
Modernist approach. The
disciplined expression in its
curves obviates a need for
surface decoration.

COMO TEA AND COFFEE SERVICE
Lino Sabattini
for Christofle 1957–60

This silver-plated French service with wicker handles has a streamlined styling which is exaggerated to give it a sense of personality. It is typical of a period in which designers began to explore new ways to invest products with meaning. From the tall, slender coffeepot to the short, stout teapot, Italian designer Sabattini combines unity of appearance with a quirky expressiveness.

CYLINDA-LINE
Arne Jacobsen
for Stelton 1967

Stainless steel became popular after 1945 initially because of shortages of silver. But by the 1960s it had developed an aesthetic associated with the light, refined elegance of Scandinavian Modern design. This classsic range of utility objects in stainless steel by Denmark's most famous modern architect, using the cylinder as its basic form, pioneered more widespread acceptance of the material.

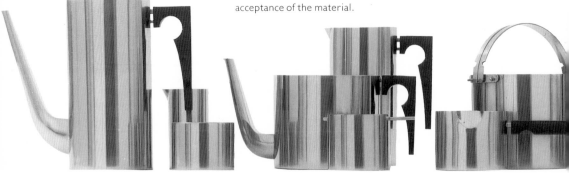

VACUUM JUG
**Erik Magnussen
for Stelton** 1977

A decade after Jacobsen's
highly acclaimed Cylinda-
Line, fellow Dane Erik
Magnussen successfully
extended the collection – no
easy feat. The vacuum flask in
stainless steel and ABS plastics
captures Jacobsen's spirit
perfectly. It opens and closes
automatically with the help of
a tilting-rocker stopper.

CONDIMENT SET
**Ettore Sottsass
for Alessi** 1978

This beautifully proportioned
condiment set for oil, vinegar,
salt and pepper with
accompanying Parmesan
cheese holder is one of
Alessi's best-selling lines. Yet
it marked Sottsass's first
experience of designing
within the constraints of
stainless steel after more
liberating plastics materials.

TEABOWL
Carsten Jørgensen
for Bodum 1984

Bodum's design director expertly balances a highly polished stainless-steel half-sphere on three legs, tops it with a conical lid and completes the full sphere with the sweep of a black plastics handle. The Teabowl is perfectly composed to be attractive to the eye and dedicated to its function. It has an unusually large tea-infuser inside and a delicate hand finish for all its surfaces, which belies the fact that it is a manufactured item, since it looks like a work of art.

FRUIT BOWL
George Sowden
for Bodum 1986

This stainless-steel bowl combines the calm grace of Swiss company Bodum's finest pieces with the decorative invention typical of Sowden's credentials as a member of the Memphis group. The patterning is achieved by a 12-stage stamping process with high-tech production techniques to create a craft effect. This work has been regarded as an attempt to establish a new Modernism.

TEA AND COFFEE SET
**Michael Graves
for Alessi** 1983

Alessi's Tea & Coffee Piazza Series, which began manufacture in 1983, produced some of the most lavish and inventive tableware of the 1980s. This silver service by American Post-Modernist Michael Graves is a miniature cityscape on a tray and became an instant classic. It has mock-ivory handles, lacquered aluminium spheres and black Bakelite feet.

TEA AND COFFEE SET
**Oscar Tusquets
for Alessi** 1983

Another range from the Tea & Coffee Piazza Series shows the extent to which everyday tableware objects have become high-style symbols, while architectural ideologies have been translated into domestic products. The Spanish architect Oscar Tusquets created the silver forms of his teapot and jug by forging two shells, welding them together and adding relief rivetting along a tilted axis. The handle is integral to one shell, while the spout is part of the other.

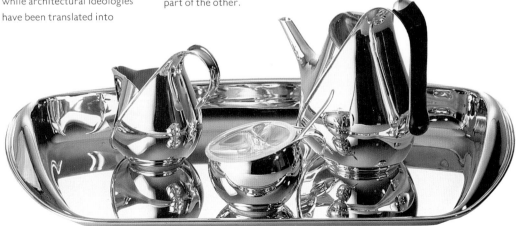

Tradition and innovation, hand and machine, art and technology have coexisted in the development of glassware during this century. Spectacular quantum leaps, such as the invention of heat-resistant glass for use in finely dimensioned all-glass tea services, have been offset by a revival of interest among contemporary Italian designers in ancient techniques first explored on the island of Murano.

Glass is a recyclable resource, made from sand and continually reused. It is also an essentially plastic medium, capable of being tempered, teased and blown into an infinite variety of forms. This versatility – an ability to create extraordinary delicate shapes which can still offer toughness and durability in use – has enabled twentieth-century designers to explore their cultural preoccupations through the style and appeal of table glass.

GREEN SODA GLASS
James Powell & Sons
c. 1880
This British-made glass service is in the Venetian style – bright, light and delicate. It captures that traditional spirit of fantasy and artistry which delights in a complexity of form. Green soda glass is today sought after by collectors.

TEA SERVICE
Ladislav Sutnar
for Schöne Stube 1931
The first all-glass tea services came on to the market in Europe following the development of heat-resistant glass at the Corning Glassworks in the USA. Thin borosilicate glass has a slow expansion rate and good chemical stability. It also offers the potential for fine dimensions in design, as demonstrated by this highly attractive example from Czechoslovakia's prestigious glassware industry. Sutnar's service achieves a pleasing form, but all-glass tableware tends to shift the visual focus from container to contents.

TEA SERVICE
Wilhelm Wagenfeld
for Jaener Glassworks
1931–4

The best known of all the pioneering heat-resistant glass tea services, Wagenfeld's range for Jena was developed at a time when the famous German designer was teaching at the Berlin Kunstochshule. It has an almost ethereal quality in its fine lines and lightweight appearance (the same glass is used to make test tubes). Note the removable glass tea-infuser in the teapot, which stands just 10 cm (4 in) high.

PRESSED-GLASS SET
Aino Aalto
for Iittala 1932

This typically Finnish glass collection was designed for a competition organized by the manufacturer but went on to win a medal at the 1933 Milan Triennale and remained in production until the end of the 1950s. Aino Aalto, wife of Finland's most famous architect Alvar Aalto, created this timeless classic.

FAZZOLETTI VASE
Paolo Venini
for Venini Glassworks
1954

Known as the 'Handkerchief Vase', this famous design exemplifies the innovative spirit of Paolo Venini (1895–1959), founder of the Venini Glass Company, who revived many traditional Venetian techniques. Designed just five years before his death, it is made by heating a sheet of *vetro a filli* glass, draping it over a form and allowing it to sag. It demands artistry and sensitivity in the making, and has often been made cheaply and badly copied.

GIGOGNE DURALEX GLASS
Saint-Gobain Glassworks 1948

This simple stacking tumbler, moulded in pressed glass, has emerged as an anonymous classic through widespread use, its distinctive faceted design is a familiar sight across Europe. Duralex was developed at the Saint-Gobain Glassworks in 1939

GLASS VASES
Nils Landberg for Orrefors 1957

The ultimate in fluidity of expression in glass, Nils Landberg's design sets the limit in the development of delicate, elongated forms. Popular during the late 1950s, it was mass produced by Orrefors, enhancing their reputation as a company sympathetic to the vision of fine artists.

VICTORIA WINE SERVICE
Frank Thrower
for Dartington Glass
1963

The English Dartington company has been much influenced by Scandinavian developments in glassware, but its best collections achieve their own presence and personality. Created by Frank Thrower, Dartington's chief designer, this robust and enduring combination of function and artistry has been one of his most successful.

DRINKING SET
Adele Melikoff
for Moser 1960s

Czech glass enjoys a reputation for its beauty and quality. This collection is typical of Moser's output, a strong, positive design well executed in Bohemian crystal. The facets that determine the look have been much imitated by lesser lines. During the 1960s glassware became a major Czech export all over the world.

BLUE STRIPE SERIES
Ulrica Hydman-Vallien for Kosta-Boda 1979

The Blue Stripe range is typically Swedish in its cool simplicity, harmony and restraint. Just as the Royal Copenhagen Porcelain Manufactory established a benchmark in 1964–5 for simple, utilitarian tableware which retained a pure Modernist grace with its Blue Line by Grethe Meyer, so Kosta-Boda, Sweden's largest and oldest glass company – it was founded in 1746 – achieved the same kind of status with Hydman-Vallien's design.

RIGEL
Marco Zanini
for Memphis Collection
1982

One of the most potent symbols of the Memphis revolution in product design, this piece by Zanini was produced in Murano glass by Toso Vetri d'Arte. It departs radically from Scandinavian-inspired cool Modernism in glassware, adopting a more exuberant approach to colour and form. In particular it combines the functional (lid and stem) with the comical: an appendage protruding on the left is seemingly added just for fun. Zanini worked closely with Memphis-guru Ettore Sottsass in the 1980s, as a member of both Memphis and Sottsass Associati.

CRYSTAL GLASS
Matteo Thun for Campari 1986

This hand-blown long drinking glass designed by Sottsass Associati member Matteo Thun clearly demonstrates the effect of the Post-Modernist movement in international design on glassware during the 1980s. A free-form variation on the traditional flute, it abandons harmony, and balance in favour of deliberate sculptural assymetry. The design was produced in a limited edition for Campari by the Austrian manufacturer J & L Lobmeyr and recalls Memphis design.

CARTOCCIO GLASSES AND VASES
Carlo Moretti 1983

Hand-blown and hand-finished, these pieces are typical of the dedication of the Carlo Moretti company, established in 1958 to maintain the ancient Murano glass tradition. Yet they also show the finesse and invention with which modern Italian designers have explored new forms. This design simulates in a delicate, controlled way a sheet of paper twisted into a scroll.

CUPOLA GLASSWARE
Mario Bellini and Michael Boehm
for Rosenthal 1987

Introduced as part of the celebrated Studio-Line, Cupola expresses a marriage of Italian flair and German order in a range characterized by a twin stem. Its more conventional clarity and harmony marks a change from the often wayward visual experiments conducted by Italian glassware designers in the mid-1980s. But then Bellini has always been the most rational talent in the hothouse of Italian design.

ORSAY GLASSWARE SET
Thomas Bastide
for Baccarat 1989

The classical architectural balance of this range with its faceted plinths reflecting light, epitomizes the quality of Baccarat crystal. The French Baccarat company was founded in 1764 and is a world leader in handmade crystal tableware, claiming to employ more award-winning craftsmen than any other factory. Chosen for its association with Bacchus, god of wine, its name indicates a dedication to the gourmet.

GLASS CENTREPIECE
Borek Sipek
for Sawaya & Moroni
1985

Just as the Victorians decorated their tables with ornate crystal *epergnes* to hold fruit and make a visual focus, so this glass vessel by Czech designer Borek Sipek updates that convention as a bold expression of mid–1980s Post-Modernism. Provocative in form, it challenges contemporary values.

There has been no more
dramatic impact
on twentieth-century
tableware than that made by
polymer technology. Plastics
have brought new and
unlimited design possibilities to plates, picnic
sets, jugs and bowls. The earliest synthetic
wares may have been cheap in image and price,
but more contemporary plastics products by such
companies as Kartell and Guzzini have become
expensive objects of desire.

Plastics have also had an influence on the
centuries-old tradition of cutlery-making,
introducing the era of disposables and
dishwasher-proof nylon and acetal handles. One
is left wondering what such notable cutlery
designers as Georg Jensen, Josef Hoffmann and
Charles Rennie Mackintosh would have achieved
if plastics had been developed in their time.

SILVER CUTLERY
Josef Hoffmann
for Wiener Werkstätte
1904
This sterling-silver cutlery range designed by one of the most important pioneers of modern design was stamped from single sheets of metal. Its absolute minimum of decoration was ridiculed at the time. One critic said that Hoffmann made 'geometry, not art'. But the collection was simply ahead of its time and typical of Hoffmann's exploration of new geometric forms in industrial design while a member of the Wiener Werkstätte.

CONTI-NENTAL PATTERN CUTLERY
Georg Jensen for Georg Jensen Silversmithy 1908

Forged and hand-hammered in polished sterling silver, the first flatware design issued by Georg Jensen is still in production today. Classically proportioned for ease of use and visual appeal, it takes as its starting point the idea that the utensil is an extension of the hand and considers comfort and ergonomics more than making the stark visual statement of the Hoffmann cutlery (opposite).

MMM FISH KNIFE AND FORK
Charles Rennie Mackintosh c. 1903

Charles Rennie Mackintosh's cutlery designs show the influence on his work of the Vienna Secessionists, who invited him to participate in their exhibition of 1900. His very individual interpretation of Art Nouveau proved to be a significant forerunner to the Modern movement. In 1984 Sabattini reissued his cutlery, making his important design legacy available to a wider contemporary audience.

SILVER CUTLERY
Sigvard Bernadotte for Georg Jensen 1939

The clear-cut, disciplined style of Count Sigvard Bernadotte, son of King Gustav Adolf VI of Sweden and brother of the Danish Queen Mother, led Georg Jensen Silversmithy into an important new phase of development in the vanguard of Scandinavian Modernism. This elegant, precise cutlery design is still in production today. Bernadotte, the archetypal aristocrat artist, joined the firm's board of directors and greatly influenced its future direction.

CORACLE PICNIC SET
Brookes & Adams
1927–32

Fitted out with a service made of Bandalasta ware, an early marbled urea-formaldehyde plastics, this English picnic hamper was a familiar sight in the 1920s and early 1930s. Bandalasta plates, teacups and saucers proved ideal for open-air touring and outdoor picnics – lightweight, unbreakable and hygienic. The famous London department store Harrods was a major supplier of Coracle picnic sets. A complete set with no pieces missing is today a rare collector's find.

MELMEX CRUET SET
A H Woodfull and John Vale
for British Industrial Plastics 1946

The design possibilities of newly-developed plastics in tableware were demonstrated by this streamlined cruet set for BIP. It was moulded initially in urea-formaldehyde and later, in the 1950s, in Melmex melamine. Highly coloured, glossy, tough and hygienic, it quickly became a familiar sight as a utilitarian object in cafés as well as school canteens.

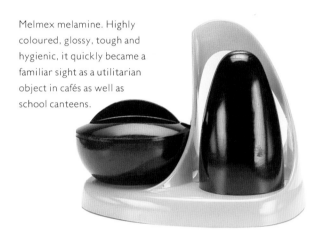

PRIDE
David Mellor
for Walker & Hall 1954

One of the most classic, enduring and popular British cutlery ranges. Mellor designed Pride in silver-plate with celluloid handles while still a student at the Royal College of Art. It was among the winners of the first ever British Design Council awards in 1957. Pride combines the modernity of its slender style and technique with an affinity with the English cutlery tradition. Now made in Mellor's own factory, it has retained its universal appeal through subtle technological updates in the design. The knife handles, for example, are today in dishwasher-proof nylon.

TAILLE CUTLERY
**Tapio Wirkkala
for Rosenthal** 1961
This organic, expressive range
in stainless steel reflects
Finnish designer Tapio
Wirkkala's background as a
sculptor. It also shows his feel
for materials – he achieves the
same harmonious dynamism
in steel as he does when
working in glass or ceramics.

ALVESTON CUTLERY
**Robert Welch for
Old Hall
Tableware** 1965
A master at bringing the
artist's tactile sensibility to
mass-produced objects,
British designer and craftsman
Robert Welch designed this
range in stainless steel. A
British Design Council award-
winner, Alveston's bold
fluency marked a milestone in
the relationship between
design and manufacture.
Welch was Old Hall's design
consultant for 28 years.

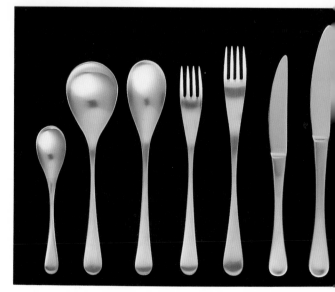

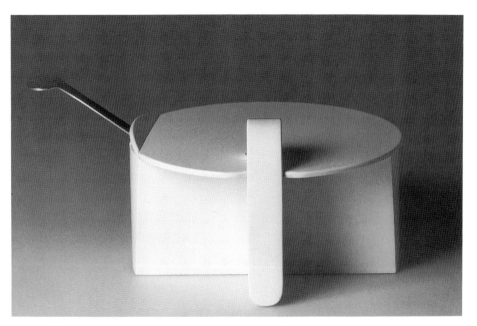

JAVA CONTAINER
Enzo Mari
for Danese 1968–9
The creative potential of plastics in tableware was seized on by Italian designers during the 1960s. This lidded container by Enzo Mari is made from two moulded pieces of melamine and has no mechanical hinge. The lid simply lifts to slot back into a vertical position over the handle, which doubles as a hinge. The container is designed to hold a variety of foodstuffs including cheese, flour and sugar

SEMI-DISPOSABLE CUTLERY
David Harman Powell
for Ekco Plastics 1970
Intended to be disposable this all-plastics cutlery range was injection-moulded in grey-tinted styrene acrylonitrile and was simply too nice to throw away!

PROVENCAL
David Mellor
for David Mellor Design
1973
One of British David Mellor's most popular designs, this cutlery shows his flair for combining contemporary styling with a sense of history. The knife blades are made of stainless steel with a high carbon content for an efficient cutting edge. The rivets are brass and the handles, originally in rosewood, are now black acetal resin. The entire range is hand-finished.

MAX I MELAMINE TABLEWARE
Massimo Vignelli
for Heller Design 1972
The versatility of plastics was brilliantly exploited in the American market by Vignelli's stacking ware for Heller. Widely regarded as the first use of melamine in a truly modern sense without reference to traditional ceramic tableware forms, the design saves space and looks good. It won a Compasso d'Oro award for its Italian designer and is now exhibited in the Museum of Modern Art in New York.

PAPILLON
**Furio Minuti
for Guzzini** 1980s

This Italian range designed in high-quality ABS, PMMA and polypropylene plastics demonstrates well the enormous advances made by synthetic tableware in less than 50 years. What started as a cheap polymer substitute for natural materials is now often the preferred medium to create upmarket objects which are beyond the pocket of many. Papillon is made in five solid colours and a clear version. Its boldness and ingenuity are particularly evident in the vacuum jug and circular plate-holder.

SALAD SERVERS
**Anna Castelli Ferrieri
for Kartell** 1976

Made of acrylic and unusually short in length, these salad servers are the closest you can get to tossing the salad with your bare hands. The curved handles enable the servers to rest on the edge of the salad bowl. Since 1976 Anna Castelli Ferrieri has been art director for the Italian Kartell company.

EAT AND DRINK CUTLERY

Ergonomi Design Group for RFSU Rehab 1980

Designed by Maria Benkton and Sven-Eric Juhlin of the Swedish group Ergonomi Design, this combination cutlery range attains the highest visual and technical standards, upgrading the concept of design for the disabled, by creating objects which are desirable to all. The designers worked with medical experts to develop a fork with serrated knife edge, the 'knork', for people with only one hand. The other implement is a 'knoon', a combined knife and spoon; it cuts up food with a twisting movement and has a deep bowl to prevent spilling.

SELTZ CUTLERY
Davide Mercatali and Paolo Pedrizzetti for Cose Casa-Industrie Casalinghi Mori 1982
This Italian range in stainless steel with nylon handles is an excellent example of the dynamism and informality of flatware in the 1980s. Its combination of bold steel forms with heat-proof synthetic handles in bright colours has an attractive Modernist feel which blurs the distinction between formal and everyday use.

SET MAGIQUE
Andrée Putman for Sasaki 1988
An intriguing, tactile design described by its leading French designer Andrée Putman as 'ideal for desk dining', this interlocking Japanese knife and fork set is light, compact and ingenious. The stainless-steel flatware is set into mat ABS plastics handles. The two utensils slide together to form a single slim, black case with exterior faceting which makes the cutlery comfortable to use.

ALVAR AALTO

Finland's master architect of
the Modern movement, Alvar
Aalto (1898–1976) created a
humanistic and democratic
visual language through the
medium of natural materials
such as brick, wood, copper
and glass.

His brand of Modernism was
more lyrical, gentle and
natural than the severe
experiments in concrete and
steel conducted elsewhere in
Europe. The simplicity and
sophistication of his designs
for plywood furniture for
Artek and glassware for Iittala
also brought recognition.

Aalto was professor of
experimental architecture at
the Massachusetts Institute of
Technology in 1946–8; and he
won the Gold Medal of the
Royal Institute of British
Architects in 1957.

CLARICE CLIFF

An Art Deco celebrity in
Britain in the 1930s, Clarice
Cliff (1899–1972) enjoyed
great popular acclaim with
her ceramic tableware designs
for the mass market.

She began work in a
Staffordshire pottery at the
age of 13, learning the craft of
freehand painting on
ceramics. A tenacious self-
educator from a humble
background, she later trained
at Burslem School of Art in
the Potteries and worked for
local manufacturer A J
Wilkinson before setting up
her own studio.

When her handpainted
Bizarre pottery, with its
brilliant colours and
geometric shapes, was
launched in 1929 it became an
instant success. So great was
the demand that she hired
young painters to follow her
style. Subsequent Cliff
pottery designs such as
Fantasque and Biarritz
enhanced her reputation.

She art directed work by
other artists commissioned by
A J Wilkinson, including
Vanessa Bell, Duncan Grant,
Laura Knight and Paul Nash,
and painted decorative
patterns created by painters
such as Graham Sutherland
onto domestic ceramics made
by Royal Staffordshire
Pottery.

Today Cliff originals and
legitimate reproductions
fetch high prices at auctions,
although outright forgeries
are plentiful.

CHRISTOPHER DRESSER

A Glasgow-born botanist who
based many of his domestic
designs on metaphors of plant
life, Christopher Dresser
(1834–1904) is widely
acknowledged as the first
industrial designer of the
modern age.

He studied at the
Government School of Design
in London from 1847 to 1854.
This brought him into contact
with Henry Cole and other
design reformers of the mid-
Victorian age, and his training
spanned the 1851 Great
Exhibition and the soul-
searching about design which
followed.

When a thriving career in
botany foundered with an
unsuccessful application for
the Chair of Botany at
University College, London,
Dresser turned to design
instead. From 1862 onwards
he prospered as a freelance
designer. Within a decade he

had built up a large practice, designing glass, ceramics, tableware and wall-decorations for popular use. He was opposed to the direct naturalistic use of plant motifs in design. Instead he adapted their formal elements of construction for manufacture.

Dresser boasted that he designed for all branches of manufacture. His metalware in particular shows a utilitarian interest in Functionalism which prefigures Bauhaus design concerns.

KAJ FRANCK

One of the best-known figures of the Scandinavian Modern movement, Finnish designer Kaj Franck (1911–89) exerted a major influence on an entire generation of designers by exploring the natural harmony of pure Functionalism.

He trained at Helsinki Institute of Industrial Art (HII) and began his career as an independent designer of lighting, furniture and textiles. Then, in 1945, he became a designer at Arabia, Finland's biggest porcelain and pottery manufacturer.

The collaboration produced some of the classic objects of Finnish design, including the **Kilta service** of 1952, which has sold 25 million pieces. The association lasted until 1973, Franck later becoming the company's art director.

During this period Franck also designed for the Nuutajärvi Glassworks from 1951 to 1973, and he was artistic director of HII from 1960 to 1968.

The recipient of numerous honours and awards, Franck also designed exhibitions, interiors, cutlery, lighting, textiles and furniture, and explored plastics. A profound and uncompromising ideologist, he believed in the sparsity of self described as 'the conscience of Finnish design'.

MICHAEL GRAVES

American architect Michael Graves (b. 1934) has exerted a major influence on international design as a leading spokesman for Post-Modernism.

His Portland Public Services Building in Oregon, completed in 1982, epitomizes his approach to decoration and symbolism in buildings. But Graves has also explored a Post-Modern world in

tableware, furniture and interiors with work for Alessi, Memphis and Sunar.

Graves trained at Harvard, rising to prominence in the 1970s. His forays into 'Kitsch' kettles and coffee sets have been heavily criticized by the architectural establishment, but his contribution to building design in the Post-Modern 1980s was incontrovertible.

HERMANN GRETSCH

An industrial designer from Stuttgart, Hermann Gretsch (1895–1950) did not attend the Bauhaus even though his work embodies many of its original ideas.

Gretsch remained in Germany when the Bauhaus

founders and other leading names in art and design fled the Third Reich, and he worked tirelessly throughout the Nazi era. He became head of Der Bund Deutscher Entwurfer (Association of German Designers), which replaced the Deutscher Werkbund when it was closed down by the Nazis.

He saw himself as a guiding light for good design in a Germany darkened by the enforced exile of so much talent, and he wrote and edited many publications on design for mass production.

His **Model 1382** porcelain tableware set, one of the great enduring classics of the twentieth century, was designed in 1931 for German manufacturer Arzberg and can be seen as the crystallization of his beliefs: a synthesis of art and industry which is simple, timeless, pure and restrained.

WALTER GROPIUS

One of the most influential figures of twentieth-century design, Walter Gropius (1883–1969) changed the nature of design thinking and education, first in Europe and then in the USA.

Gropius was born in Berlin. His family were architects and he studied architecture in Berlin and Munich before working from 1908 to 1910 for Peter Behrens in the design office at AEG (Allgemeine Elektrizitäts-Gesellschaft).

In 1910 Gropius formed his own architectural practice with Adolf Meyer. Their partnership produced a series of famous buildings, including the exhibition centre for the Deutscher Werkbund in 1914. But Gropius is best known for founding the Bauhaus in 1919, a school of art and design whose impact on the material world is still felt today.

Between 1919 and 1928 his unprecedented experiment in design education at the Bauhaus shaped the ideology and direction of Modernism.

Although the school was organized into a series of craft workshops, its teaching has profoundly affected mass production.

Alarmed by the rise of Nazism, Gropius fled from his native Germany in the 1930s, first to England and then to the USA where, in 1937, he became a professor in the Graduate School of Design at Harvard.

At Harvard, Gropius set about transforming American architectural education. His persuasive belief in Functionalism made its mark on US design via his work as a teacher and as a consultant. In 1945 he founded TAC (The Architects' Collaborative) and his range extended from buildings to domestic products, including the memorable **1968 TAC I tea set** for Rosenthal.

Gropius's design legacy is all-embracing, ranging from clean, functional aesthetics to a much-imitated team approach in the design process.

JOSEF HOFFMANN

Austrian architect and designer Josef Hoffmann (1870–1956) was one of Europe's most important pioneers of modern design in the early years of this century.

He studied under Otto Wagner, the father of twentieth-century Austrian architecture, at the Vienna Academy, and entered his studio in 1896. The following year Hoffmann joined a radical group of architects, designers, painters and sculptors in the self-styled Vienna Secession. Their aim was to develop a bold new style in reaction to the stranglehold of traditional Austrian design.

In 1899 Hoffmann became professor of the School of Applied Arts in Vienna, a position he was to hold until 1941. As his influence grew, he founded, in 1903, the Wiener Werkstätte (Vienna Workshops), with Kolo Moser and Fritz Wärndorfer.

Inspired by C R Ashbee's Guild of Handicrafts in Britain, they continued for 30 years. It was here that Hoffmann's famous metal cutlery and tray designs were executed, revealing an interpretation of Art Nouveau which featured repeating geometric motifs and grid patterns rather than the floral decoration more typical of the style.

Although his greatest work was completed prior to World War One, Hoffmann's ideas – documented in a series of manifestos – have continued to influence designers. In particular his pursuit of 'good, simple things for the home' inspired successive generations of utilitarian domestic tableware.

MARTIN HUNT

British designer Martin Hunt (b. 1932) studied pottery at Gloucestershire College of Art and the Royal College of Art. In 1966 he co-founded the Queensberry Hunt Design Group with David Queensberry, a notable ceramics designer.

This practice has become renowned for designing many innovative and sophisticated tableware ranges. In particular Hunt has worked for Wedgwood, Hornsea Pottery and Royal Doulton in Great Britain, and Rosenthal and Thomas in Germany.

Between 1976 and 1986 Hunt was head of the Glass Department at the Royal College of Art. In 1983 he was elected to the Faculty of Royal Designers for Industry.

Martin Hunt has won no less than four British Design Council awards, three of them

for Hornsea Pottery tableware pieces. Recent work for the Tournée porcelain range by Thomas has confirmed his reputation as a master of form and one of Europe's most important ceramics designers.

ARNE JACOBSEN

A celebrated master of Functionalist Scandinavian design, Arne Jacobsen (1902–71) graduated in architecture from Copenhagen's Academy of Fine Arts in 1927 before beginning an extraordinary career which encompassed not only buildings but practically every material object which furnishes them.

Jacobsen's vision was shaped by a love of nature and a strong interest in painting and sculpture. Thus he was able, in the 1950s, to give the austere values of the Modern Movement a more naturalistic, refined and elegant character.

He took as much interest in the details of a physical environment – the lighting, tableware, cutlery, textiles and furniture – as in the outward form of the building. St Catherine's College in Oxford remains a fine example of his total, multidisciplinary approach.

Jacobsen designed a series of moulded wood-laminate chairs for Fritz Hansen in the 1950s which have become classics, and his **1967 Cylinda-line** stainless-steel tableware range for Stelton endures to this day as a testament to his quest for control and perfection.

GEORG JENSEN

The founder of one of the world's most famous silverware companies, Georg Jensen (1866–1935) grew up in a rural setting just north of Copenhagen. Despite a humble upbringing, his ambition was to become a sculptor.

Apprenticed to a goldsmith at the age of 14, he used his metalworking craft to support his endeavours in sculpture and ceramics. Although he attracted international attention as a ceramicist, he was unable to support himself and when almost 40 decided to concentrate on metalwork, opening the Georg Jensen Silversmithy in April 1904.

Many leading artists, sculptors and designers have been associated with the Georg Jensen company over the years, among them Johan Rohde, Sigvard Bernadotte, Henning Koppel, Harald Nielsen and Kay Bojesen.

RAYMOND LOEWY

A flamboyant showman who became one of the pioneers of the modern international design-consultant business, Raymond Loewy (1893–1986) was born in Paris.

He went to New York in 1919, having been demobbed from the French Army, and gained work as a fashion illustrator and display designer. The concept of the industrial designer in the USA was not established at that time, but a turning point came in 1929 when Loewy was given five days to redesign a Gestetner duplicating machine. The result – a clever styling exercise – has passed into design folklore as one of the milestones of design in industry.

Together with Henry Dreyfuss, Walter Dorwin Teague and Norman Bel Geddes, Loewy became a pioneer of design consultancy. This group is associated with the rise of styling in the 1930s, a conscious manipulation of form sneered at by the purists but which nevertheless ensured that design became an integral part of the production process.

Loewy was the most successful and uninhibited of his peer group: by 1946 he had 75 international clients in a wide range of industries who paid him both retainers and royalties.

Much of his work became icons of contemporary consumer life – from the Coca-Cola bottle and Lucky Strike pack to the Studebaker car and tableware for Rosenthal. His 1951 autobiography, *Never Leave Well Enough Alone*, added to the carefully cultivated image of living legend.

He was the first designer to make the cover of *Time*, and

his influence extended into the Kennedy era: Loewy designed the interiors for JFK's personal Boeing 707 and, while working for NASA from 1967–72, was involved in the design of spaceship interiors. A London consultancy bearing Loewy's name survives today.

DAVID MELLOR

British cutlery and industrial designer David Mellor was born in 1930 in Sheffield, a city historically associated with steel and cutlery. In 1954, he opened his first studio–factory–workshop there and has since become one of the most important craftsmen-designers of cutlery and kitchenware in Europe.

His **Pride** cutlery of 1954 won a British Design Council Award and he has subsequently developed careers in design, manufacturing and retailing. Mellor runs shops in London and Manchester and has recently opened an architecturally innovative circular factory at Hathersage near Sheffield. He received the coveted Medal for Industrial Design from the British Chartered Society of Designers in 1988.

GIO PONTI

Few personalities have exerted such an influence over twentieth-century Italian design as Gio Ponti (1891–1979).

In 1927 he founded the architecture and design magazine *Domus*, which governs opinion and makes reputations to this day; he was a moving force behind the Milan Triennale exhibition, which put Italian designers on the world stage; and he also designed the Pirelli office tower, in Milan in 1956, a classic of post-war European architecture.

As professor of Architecture at Milan Polytechnic, Ponti was in a position to use his skills as a propagandist to influence an entire generation of Italian designers. He helped to

establish the prestigious Compasso d'Oro awards for Italian industrial design and he undertook commissions in furniture, sanitaryware and tableware, working for such names as Cassina and Arflex.

JEAN PUIFORCAT

French designer Jean Puiforcat (1897–1945) was the outstanding Art Deco sculptor–silversmith of the 1920s and 1930s.

After serving with the French Army during World War One, and winning a *Croix de Guerre* in 1918, Puiforcat went to work in his father's silver workshop. By 1925 he was acclaimed as a master of his craft: at the Paris Exhibition that year, a special area was devoted to his designs.

Puiforcat's work for both French and English companies was based on the harmony and geometry of ancient Greek mathematicians, rather than on elaborate flourishes – the usual stock in trade to satisfy wealthy patrons of silversmithing. Puiforcat has been described as a specialist in 'costly simplicity', but the enduring appeal of his finest pieces indicates a more profound design sensibility.

ETTORE SOTTSASS

The man who changed the face of international design in 1981 with the first Memphis collection of ceramics, chairs and lights, Ettore Sottsass (b. 1917 in Austria) has proved to be one of the most enigmatic and influential talents of post-war Italian design.

Trained as an architect at Turin Polytechnic, Sottsass ran his own practice in the city before working as a consultant to the electronics division of Olivetti during the 1950s and 1960s.

Later he became associated with prestigious manufacturers such as Alessi and Poltronova, for whom he designed tableware and furniture, as well as with avant-garde design groups in Milan.

Sottsass's reputation as a radical was confirmed when he shattered the refined Modernist consensus of Milanese high design with his wild, garish and controversial Memphis collections. This placed him firmly at the head of the international Post-Modernist movement, despite his protests that Memphis collections were not supposed to be taken seriously.

Today Sottsass works with architects Aldo Cibic, Matteo Thun and Marco Zanini in his Sottsass Associati design consultancy formed in 1980.

GEORGE SOWDEN

British designer George Sowden (b. 1942) studied architecture at Cheltenham College of Art between 1960 and 1968. He moved to Milan in 1970 and joined the Olivetti studio run by Ettore Sottsass, where he developed designs for information technology.

In 1981, Sowden was among the founder members of Sottsass's Memphis movement, creating furniture and objects for successive Memphis collections. He has subsequently founded his own consultancy, working either independently or in partnership with Nathalie du Pasquier.

Sowden's tableware designs for Bodum have been widely acclaimed in Europe as combining the best functional elements of Modernism with the best decorative elements of Memphis.

OSCAR TUSQUETS

Born in Barcelona in 1941, Spanish architect Oscar Tusquets studied architecture in his native city before founding the Studio PER with architects Pep Bonet, Cristian Cirici and Lluis Clotet in 1965.

Until 1984 Tusquets worked jointly with Clotet on most projects. In addition to architecture, Tusquets has been a consultant designer to many companies, and a guest lecturer and professor in Europe and the USA. From 1975 until 1976 he was also professor in charge of projects at the *Escuela Tecnica Superior de Arquitectura*, where he was educated.

In the early 1980s he joined a select group of 11 international architects commissioned by Italian manufacturer Alessi to design prestigious tea and coffee sets.

TAPIO WIRKKALA

Finnish designer Tapio Wirkkala (1915–85) is closely associated, like Kaj Franck, with the ideals of the Scandinavian Modern movement. In particular, his glassware designs for the Finnish glass company Iittala from 1947 onwards greatly influenced the world of glass and led to a succession of international honours and awards.

Wirkkala played a large part in Scandinavian design's eruption onto the world stage in the 1950s. He designed Finland's stand for the Milan Triennale in 1954, and became a highly popular figure in table glass.

In 1955 he joined Raymond Loewy, another charismatic design name, and went on to work for Rosenthal and Venini.

RUSSEL WRIGHT

The most outstanding American tableware designer of his generation, Russel Wright (1904–76) was a household name to US consumers in the 1940s and 1950s. Born into a family of Quakers, the restrained and puritan values in his designs equated him in some eyes with the Bauhaus doctrine.

At his father's insistence, Wright entered Princeton to study law but did not last the course. He went to study sculpture at the Art Students' League in New York and pursued a career in the theatre. His earliest success was with a series of life-size caricature masks of celebrities, which received an enthusiastic note in *Vogue*, but Wright soon began to explore industrial design.

His first table-serving pieces were in spun aluminium and, encouraged by his wife's flair for public relations, he began to develop successful new ways to market them during the Great Depression of the early 1930s. Although a shy man, he nevertheless signed his pieces and promoted them through personal appearances. He even persuaded Macy's, the

New York store, to use his name on advertisements, the first time a designer enjoyed such a privilege.

Sterling-silver cutlery followed in 1933, and Wright also developed new furniture lines under the banner Modern Living. But his outstanding success was the ceramic **American Modern** dinnerware collection, a US design classic made by the Steubenville Pottery for 20 years from 1939.

American Modern captured the spirit of the age and Wright himself became a celebrity. After World War Two he took his gift for natural lines and simple, popular shapes into the field of plastics, designing the first mass-market range in melamine, the Residential dinnerware collection of 1953.

At the peak of his fame Wright retreated to his country estate. He was devastated by the failure of a home-furnishings venture he had designed called American Way but more particularly by the death of his wife in 1952. He worked only sporadically, but his reputation as one of the most important figures in twentieth-century American design and consumer culture was already secure.

DATE DUE